WEATHER WATCH

J
551.1
Wya
Wyatt
Weather watch

6913024
8.95

Weather Watch

WRITTEN BY VALERIE WYATT

Illustrated by Pat Cupples

ADDISON-WESLEY PUBLISHING COMPANY

Reading, Massachusetts   Menlo Park, California   New York
Don Mills, Ontario   Wokingham, England   Amsterdam   Bonn
Sydney   Singapore   Tokyo   Madrid   San Juan
Paris   Seoul   Milan   Mexico City   Taipei

First published in Canada by Kids Can Press, Ltd., Toronto, Ontario

Library of Congress Cataloging-in-Publication Data

Wyatt, Valerie.
    Weather watch / written by Valerie Wyatt : illustrated by Pat Cupples.
       p. cm.
    Summary: Explains what makes weather, why it changes so often, and how it affects humans and animals. Includes activities and experiments.
    ISBN 0-201-15404-8
    1. Weather—Juvenile literature. 2. Weather—Experiments—Juvenile literature. 3. Weather forecasting—Juvenile literature.
[1. Weather. 2. Weather—Experiments. 3. Experiments.]
I. Cupples, Pat, ill.   II. Title.
QC981.3.W93   1990   551.5—dc20   90-491

Edited by Charis Wahl
Designed by Nancy Ruth Jackson
Cover illustration by Pat Cupples
Set by Compeer Typographic Services Limited

8 9 10 11 12-CRS-9998979695
Eighth printing, February 1995

CONTENTS

ACKNOWLEDGEMENTS

I am grateful to many people for their help and advice on this book. Ed Truhlar at the Atmospheric Environment Service of Environment Canada went over the initial draft with a fine-toothed comb. He caught many errors and made numerous valuable suggestions. Other consultants include: Elaine Christens; Dr. C.R. Harrington at the National Museum of Natural Sciences; Dr. Larry Licht at York University; Ken Lister at the Royal Ontario Museum; Philip Mozel at the McLaughlin Planetarium; Dr. R.S. Schemenauer; and Dr. Audrey Karlinsky. Thank you all.

Weather Watch had several "midwives." My publishers and friends Ricky Englander and Valerie Hussey at Kids Can Press gave me unqualified support and encouragement during the writing. Charis Wahl made me laugh all the way through the editing process and improved the book immeasurably. And Nancy Jackson and Pat Cupples made the visuals sparkle.

Finally I would like to thank my most important advisor, my husband, Larry MacDonald, who is an amateur weather nut. Without his help, advice and weather know-how, this book would still be in the word processor.

In memory of Fred Wyatt
who spread sunshine all around him

WEATHER WONDERS

If you like flying kites, splashing through puddles, throwing snowballs or walking in a fog, this book is for you. It's about the wonderful, sometimes weird weather that swirls around us every minute of every day.

The weather is a constantly changing show. Blue skies and sunshine give way to a curtain of snow so thick you can barely see through it. Glowing rainbows chase thunder and lightning. And clouds shaped like dragons turn into lambs right before your eyes.

But the weather is more than entertainment. Weather shapes the way you live.

Take a look around the room you're sitting in. Is there glass in the windows? What about a heater or radiator? You wouldn't need window glass or heat if you lived in the tropics. You'd want the air to blow through and cool things off. How your house is built depends on the weather where you live.

What about your clothes? Sure, you may have picked them because they're all the rage. But you wouldn't pull on your fabulous fish shorts in a blizzard no matter how much you loved them.

And then there's food. Without weather, there'd be no peanuts for your peanut butter — and no bread to put it on. Plants need sun and rain to grow. Even a hamburger needs sun and rain. After all, cows eat plants.

Without weather, Earth would be more like the moon — a cold, lifeless boulder barrelling through space.

In this book you'll find out what makes Earth's weather and why it changes so often. You'll also find out how to make weather right in the kitchen. But before you start reading, why not go outside and enjoy the weather:

- catch a snowflake (page 40)
- make a rainbow in the backyard (page 36)
- preserve a raindrop (page 32)

Then turn the page and find out more about what makes the weather.

WEATHER AND YOU

When you woke up this morning did you look outside to check out the weather? Chances are you did. You may not think about it a lot, but you pay attention to the weather. If you didn't, you might find yourself wearing a parka on a hot summer day or grabbing your skateboard when what you really need is your skates.

Read on to find out some surprising ways that weather affects your body, the way you live and the animals you share this planet with.

SHIVERING AND SWEATING

If you visited Vostok, Antarctica, and then went to Al'Aziziyah, Libya, your body would be in for a shock. Your trip would have taken you from the coldest place on earth to the hottest.

Adjusting to weather extremes is tough on the human body. Your body feels most comfortable when the temperature is around 25°C (77°F). If it's hotter or colder, your body automatically starts to adjust.

What would happen if you landed in Vostok, Antarctica, where the temperature has dropped as low as −88.3°C (−126.9°F)? You'd shiver. When you shiver all your muscles tighten; these muscle contractions give off heat, which warms you. Contracting other muscles will make you warmer, too. In other words, try some exercise! In Vostok you might want to shovel snow: there's bound to be plenty of it.

If you stayed in Vostok for a few weeks, your internal temperature would drop slightly, so that your body wouldn't have to work so hard to keep you warm. And, so you'd lose as little heat as possible, your blood would take deep routes through your body, instead of travelling near the surface of your skin as it does in summer. These changes would happen automatically to make you more comfortable.

One way to warm up would be to travel to Al'Aziziyah, Libya. In Al'Aziziyah, the temperature has soared as high as 58°C (136.4°F). First thing you'd notice would be the sweat pouring off you.

The water in sweat comes from your blood. It carries heat from your body out through your sweat glands. A patch of skin about the size of a quarter has about 300 sweat glands.

How much you perspire depends on how big you are and how hot it is. A large man can sweat as much as 19 L (17 quarts) of water a day. Even on a cool day, you pump out enough sweat to fill a small glass.

Sweating is an effective way to cool off, except when it's humid. On humid days, there's already so much moisture in the air that

there's no room left for your sweat. Instead of evaporating to cool you off, the perspiration just clings to your skin, making you feel hot and sticky—and crabby.

You probably wouldn't want to live in Vostok or Al'Aziziyah. But heat or cold anywhere will bring out the body's defences, and before long you'll be shivering or sweating your way to comfort. For more tips on how to take the brrrr and gasp out of the weather, see the box on this page.

HOT AND COLD SURVIVAL TIPS

Knocked out by the heat? Try these tips:

- drink plenty of fluids to replace the water you're losing in sweat. The best drink is cold water.
- don't exercise strenuously. Remember—working your muscles produces heat, the last thing you need.
- wear a sun hat and sunscreen when you're going to be out in the sun.

To stay warm in cold weather, try these tips:

- wear a warm hat outside. You lose as much as half your body heat through your head.
- give your body time to adjust to the cold. After a while your body temperature will drop slightly.
- put lots of protein-rich fuel (food) in your furnace (body) to make more internal heat.
- stay dry. Change wet clothing right away.

13

CHANGING THE WEATHER

Suppose the temperature hits 38°C (100°F) and you're wiped out by the heat. You've downed so much lemonade that your lips feel permanently puckered, and all you want to do is share a bath with some ice cubes. What you need is a COLD button you can push to make some snow fall.

People have dreamed of changing the weather — or at least controlling it — for thousands of years. Long ago people invented weather gods and prayed to them for good weather. There were wind and rain gods, sun gods and lots of thunder and lightning gods. The god you see here is Lei Chen Tzu, the ancient Chinese version of a thunder and lightning god.

In the Nordic legends from Scandinavia, the thunder god was Thor. He was a superstrong and superpowerful god (although not always supersmart) who gave us the name for a day of the week — Thor's day or Thursday. Even Santa's reindeer had connections with thunder and lightning. In German *Donner* means thunder and *Blitzen* means lightning.

As people began to learn how the weather works, they realized that the gods had no control over it. Nor, it seemed, did anyone else.

Most people forgot about changing the weather. But a few dreamers didn't give up.

A lot of inventive thinking went into how to make artificial rain. One inventor even tried sending a bomb into the clouds and setting it off. In the 1940s two Americans finally made rain fall by sprinkling clouds with tiny bits of dry ice. They knew that many clouds won't produce rain unless there are minute ice crystals in them. They simply helped the clouds along by supplying the ice. This "cloud seeding" made a big splash in Paris. It produced a downpour of 82 000 tonnes (90 000 tons) of water, enough to fill about 43 Olympic swimming pools. Cloud seeding is still done today. But instead of tiny ice crystals, a chemical called silver iodide is used.

Sometimes people accidentally change the weather — and not always for the better. For

Notice the fire around Lei Chen Tzu's head and the drum at his side—necessary equipment for a thunder and lightning god.

example, by building cities, people have created "heat islands." City buildings trap the heat. And the winds channelled by big buildings make some city streets so blustery that pedestrians can be blown off their feet. City weather is also wetter, foggier and smoggier than the surrounding countryside.

Even bigger weather changes are happening because of the pollution we pump into the atmosphere. For more about what pollution is doing to the weather, see pages 52–57.

Looking (and Feeling) Cool

You may not be able to push a COLD button on a hot summer day, but you can dress so you not only look but feel cool. But . . . hmmmm . . . what to wear? Try this experiment for some hot suggestions, then turn the page to find out about some really cool clothes.

You'll need:

○ *2 identical glasses*
○ *black paper big enough to wrap around one glass*
○ *white paper big enough to wrap around the other glass*
○ *tape*
○ *an outside weather thermometer*

1. Wrap the black paper around one glass as shown. Try to cover as much of the glass as possible. Tape the paper in place.
2. Cover the other glass with the white paper, and tape it too.
3. Fill both glasses with water that's about room temperature.
4. Set both jars on a sunny windowsill.
5. Come back in an hour or so and use the thermometer to see if the water in both glasses is still the same temperature.

On a hot summer day, which would you rather wear: light or dark colours?

WEATHER WEAR

This outfit may not look too cool to you, but it is. In fact every piece of clothing you see here was specially designed to keep a person cool—or from feeling cool.

Need to keep the sun off your head? Try a leaf umbrella. Grow-your-own umbrellas were common thousands of years before people started making them.

Snow goggles like these were worn as long as 2000 years ago by the Inuit in the Arctic to save their eyes from snow blindness. To make a pair of your own, follow the instructions on the next page.

Feeling chilly? Slip on Inuit leggings made from strips of twisted rabbit skin woven together. Although you could stick your fingers right through holes in these leggings, they hold in body heat and keep you warm.

Fifteen-year-old Chester Greenwood invented ear-muffs in 1873. He asked his grandma to sew chunks of beaver fur onto the ends of a wire because he was sick of freezing his ears while skating.

A bamboo undershirt may not sound very comfy, but the ancient Japanese found that it let lots of air through and kept its wearer cool.

16

FAR-OUT SUNGLASSES

Far-North sunglasses might be a better name for these cool-looking shades. The thin slits reduce the amount of light that reaches the eye and so protect the eye from glare. Inuit snow goggles like these were made of bark or bones. This version cheats a little: it uses egg cartons instead.

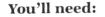

You'll need:

o *the bottom of an egg carton*
o *scissors*
o *string*
o *paints and things to paste on for decoration*
o *glue*

1. Cut out two adjoining egg cups from the carton.

2. Use the end of a scissor blade to poke a thin slit across each egg cup.

3. Cut out a triangle between the cups so the glasses will fit your nose.

4. Poke holes at the edge of the glasses and tie on the thread.

5. Paint and glue on decorations to make your far-out glasses look really cool.

6. Tie them on.

HOT DOGS: HOW WEATHER AFFECTS FIDO

Next time the weather turns hot, watch a dog. Like you, Fido is affected by the weather. Instead of romping around energetically, Fido will sprawl out and only move if it's absolutely necessary. When he does move, he does it *s-l-o-w-l-y*. Fido also pants a lot. This is his way of getting rid of excess body heat, because dogs don't sweat.

Pets may feel uncomfortable when the weather gets steamy or chilly, but for wild animals weather is much more serious. Bad weather can dry up their sources of food, wash away their homes, freeze their young and threaten their survival.

Cold-blooded animals are particularly sensitive to the weather. Here's an experiment to see how hot and cold weather affects one cold-blooded animal—the common housefly. Don't worry. This won't hurt a fly.

You'll need:

o *a fly*
o *a jar with a lid*
o *a hammer*
o *a nail*

1. Hammer tiny holes in the lid of the jar so that the fly has some air to breathe.
2. Put the fly in the jar and screw on the lid.
3. Put the jar in the fridge for half an hour and then take it out. What's happened to the fly? Is it moving slower or faster than it used to? Free the fly when you're done.

Insects and other cold-blooded animals such as snakes and frogs depend on the weather for their body heat. (We mammals have a body heat that stays more or less steady no matter what it's like outside.) Cold-blooded animals don't control their own body-heating system; it goes up and down with the outside temperature. On cold days, they're cold; their muscles don't warm up enough to move much. On warm days, they're warm. Their muscles heat up and they can move around as they wish.

This is why you'll often see snakes and lizards basking in the sun; they're soaking up the sun's warmth to raise their body temperatures so that they can get moving.

What if mammals were like cold-blooded animals and had to depend on the weather to warm up their muscles? For one thing you wouldn't have to worry about walking the dog on a cold winter morning. The dog wouldn't be able to move. And neither would you.

Imagine if all the babies born in winter were boys and all those born in summer were girls! Sound crazy? Not to many turtles, lizards and alligators. Their babies hatch from eggs, and whether they become male or female depends on how hot or cold it is when they are developing inside the eggs. Many turtles have female babies if the temperature is hot and males if the temperature is cooler. With alligator babies it works the other way around.

Sea turtle babies seem to have built-in thermometers. If they hatch when it's hot, they don't move. They stay in their underground nests in the sand until it cools down. The newly hatched turtles have to crawl across a sandy beach to get to the ocean. If they tried to reach the water during the hot daylight hours, they might cook on the sand. They'd also be easy for hungry predators to catch. By waiting until the temperature (and night) falls they have a better chance of surviving their dangerous journey across the beach.

Cold weather slows down some frogs and salamanders so much that it can take them up to nine times longer to hatch from their eggs. The Northwestern salamander may hatch missing a few vertebrae in its backbone if the weather is too cool while it's developing in the egg.

PUT YOUR HEAD IN THE CLOUDS

What would life be like without clouds? Very dry and very boring. Clouds are airborne water tanks; they hold and release the water we need to survive. And clouds make the weather more interesting. They form and fade above our heads in an ever-changing spectacle. In this section you'll make a cloud in a bottle and get some tips on cloud watching. You'll even find out how to make lightning in your mouth!

MAKE A CLOUD IN A BOTTLE

If you think clouds only happen up in the sky, guess again. You can make a mini-cloud right in your kitchen.

You'll need:

○ a candle
○ a match
○ a 2-L (2-quart) glass jug
  with a narrow neck

1. Turn the jug upside down and hold a lit match inside the mouth of the jug for about 5 seconds. Ask an adult to help you with this.

2. Wait until the glass cools, then put your mouth over the mouth of the jug so that it's completely covered. Blow hard and try to force some air into the jug.

3. Watch what happens inside the jug when you take your mouth away.

Your jug cloud was made with the same "ingredients" and in the same way as a real cloud.

You supplied the first ingredient when you let the candle burn into the bottle: soot. The soot particles are so small that you can't see them. But without particles of some kind, no cloud will form.

There are lots of tiny particles in the air around you. They're made up of forest-fire soot, volcanic dust, bits of rock, pollen from flowers and salt from the oceans. Exhaust from cars and smoke from industries add man-made particles. These airborne particles are so tiny they make the dust in your home look like monsters.

The second ingredient you need to make a cloud is warm, moist air. When you made the cloud in a jug, your breath provided the wet air. On a cold day you can see the water vapour in your breath. Breathe on a cold window—the water will stick to the glass.

But where's the water in the sky? You can't see it, but it's there. On a summer day there are about eight milk cartons full of water in a chunk of air that would fill a classroom. This water has been "sucked" out of lakes, rivers and oceans by a process called "evaporation."

Now that you've got the two cloud ingredients—tiny particles and moist air—how do you turn them into a cloud?

You cool them. The air in the jug cooled suddenly when you took your mouth away. The air in the sky cools as it rises. When air is cooled, some of the water vapour gets "squeezed" out, or "condenses." (Try the challenge on this page and see for yourself.) It turns back to a liquid on the tiny particles in the air. The result: tiny water droplets. A cloud is just a huge crowd of tiny water droplets.

THE WATER-HOLDING CHALLENGE

Which holds more water vapour, warm air or cold air? Try this experiment to find out.

You'll need:

o *two clean, self-sealing plastic bags*
o *a freezer*

1. Scoop some air into both plastic bags by opening them and waving them through the air. Seal the bags tightly shut.

2. Put one bag in the freezer and wait for five minutes. Leave the other bag of air out on the table.

3. After five minutes, take the bag in the freezer out. Open it, blow into it and quickly seal it shut. Could the air in the bag hold all the water vapour, or did some of it condense and stick to the sides of the bag?

4. Do the same thing with the bag that hasn't been in the freezer. How much water vapour condensed this time? Which bag had more condensation—the warm bag or the cold bag? Which can hold more water vapour—warm air or cold air?

MAKE YOUR OWN THUNDERSTORM

Most of the time clouds are pretty peaceful. But every once in a while, the clouds turn dark and—*KABOOM!*—a thunderstorm puts on a show. What makes thunder and lightning? Here are two experiments that'll help you find out.

LIGHTNING IN YOUR MOUTH

Your mouth is probably the last place you'd expect to find lightning. But you can make a lightning-like flash just by chewing.

You'll need:

o *wintergreen-flavoured Life Savers (candies)*
o *a mirror*

1. Go into a dark room and wait until your eyes have adjusted to the darkness.
2. Chew on a couple of wintergreen Life Savers with your mouth open, even though it's rude. Watch what happens in a mirror.

Crushing the Life Savers with your teeth rips apart the sugar crystals and creates mini-islands of candy that have different electrical charges. A spark of electricity leaps between the differently charged candy chunks and the wintergreen oil helps you see it.

Real lightning happens much the same way, only there's nothing to crush. The earth and the thunder-cloud develop different electrical charges. Electricity leaps between the two, much like the spark of light did in your mouth.

First a "leader" zaps out of the cloud and carves out a hot pathway in the air. Then electricity flows up this path from the ground.

This is called the "return stroke." Usually there are a couple of up and down strokes in the same lightning bolt; one record breaker had 26 strokes. All of this happens so fast that your brain thinks it has only seen one flash.

THUNDER IN A POP CAN

As the air in the lightning stroke heats up, it takes up more room. How does this cause thunder? See (and hear) for yourself.

You'll need:

o *a full unopened can of pop*

1. Listen as you open the pop can.
Before you open the can, the air inside is held under pressure. When you open the can, you release the pressure and the air expands. It's the expanding air that makes the sound, just as expanding air makes thunder rumble. The faster the air expands, the noisier the sound.

Sharp cracks of thunder mean the lightning is nearby. Low, rumbly thunder usually means it's farther away. You can tell how far away a lightning bolt is by counting the seconds between the lightning and the thunder and dividing by five. For example, if you see lightning and count ten seconds before you hear thunder, divide ten by five. The answer (two) is the number of miles between you and the lightning. If you see a flash and hear a sharp crack at the same time, the lightning is right above you!

DID YOU KNOW THAT . . .

- As you read this, there are 2000 thunderstorms raging in the world.
- A single lightning bolt contains enough electricity to power an average home for about two weeks.
- There are 100 lightning flashes every second of every day.
- Lightning bolts can be as long as several miles. Many are no thicker than a finger.

LIGHTNING SOMETIMES STRIKES TWICE

Lightning can even hit the same person twice. Just ask Roy "Dooms" Sullivan of Virginia. He's been hit by lightning seven times.

Toronto's CN Tower, the tallest free-standing structure in the world, is struck by lightning about 65 times a year.

SOME CLOUDS TO WATCH FOR

Next time you're out for a walk, do some cloud watching. Amazing things happen up there in the sky. For example, a cumulus cloud (the kind that looks like cauliflower) can reach a height of 20 km (12 miles). That's some cauliflower. Read on to find out more about some of the clouds you'll see.

NAME THAT CLOUD

Clouds have "family" names just like you and your friends do. The three main families are cumulus, stratus and cirrus. **Cumulus clouds** *form when blobs of warm, moist air float upwards from the Earth's surface.* **Stratus clouds** *form when a layer of warm, moist air rises or when it's forced up by a layer of air moving in below it.* **Cirrus clouds** *are bunches of ice crystals. They get their comma-like tails when some of the ice crystals in them start to fall. To see where to look for them in the sky, turn to page 90.*

Airplane clouds

Airliners make high-level cirrus clouds when the water vapour from their engines freezes into ice crystals. These clouds look like long white tails in the sky. You can use these "contrails" to help you forecast the weather. If a contrail vanishes quickly, the weather will be good. If it lingers, you may be in for a storm.

N(ice) clouds

Most clouds are made up of water vapour, but not high-flying cirrus clouds. They look like wispy commas and are so high up in the sky that the water in them freezes. So cirrus clouds are filled with ice crystals, not water. It's a good thing they never produce rain or . . . ouch!

Big pancakes

The flat layers of clouds that seem to cover the whole sky are called stratus clouds. They're usually not more than about 1 km (3/5 of a mile) thick, but what they lack in height, they make up for in s-p-r-e-a-d. A single stratus cloud can spread out big enough to cover all of Montana and Wyoming.

Rain? Rain?

The tiny droplets of water vapour that make up clouds are too lightweight to fall. You could fit about 15 million of them in a raindrop. How big is a raindrop? Turn to page 32 to find out.

Ground clouds

You can put your head in the clouds (and the rest of your body, too) next time it's foggy. Fog is just a stratus cloud on the ground. Take a walk in the fog and notice how wet you get. That's because the water droplets in the cloud stick to you.

IT'S RAINING, IT'S SNOWING

Rain and snow fall on your head all the time. But how much do you really know about them? Try this quiz and find out.

1. An average-sized raindrop is
   a. this big → ←
   b. this big → ←
   c. or this big → ←
2. Raindrops are shaped like
   a. tear-drops
   b. mushrooms
   c. hamburgers
3. Snowflakes are shaped
   a. like this
   b. like this
   c. or this

Stumped? Read on to find out the answers to these and other snow and rain questions.

BE A RAINMAKER

Making rain is a breeze . . . er, a snap. All you need is water and a way to heat and cool it. Try this recipe for kitchen rain. Then compare your rain making with the real thing.

You'll need:

o *an electric kettle or an ordinary kettle and permission to use the stove*
o *a tin can filled with ice*
o *tongs for holding the tin can*

1. Half fill the kettle with water. Plug it in, if it's electric. For ordinary kettles, turn a burner on the stove to High and put the kettle on it.

2. When the kettle is boiling and steam is rising from it, you're ready to make rain. Use the tongs to hold the ice-filled can above the steam. You won't need an umbrella, but you will see a few drops of home-made rain fall from the bottom of the can.

Rain in nature is made much the same way. Water in lakes, rivers and oceans is warmed by the sun, evaporates and rises. Then it cools, condenses and falls back to earth as rain.

What happens inside a cloud to make it rain? In some clouds, water droplets freeze onto tiny ice crystals floating in the cold upper part of the cloud. As more and more water droplets stick to the ice crystals, they get heavier and heavier and start to fall. On their way down, they pick up still more water and get bigger.

If these ice crystals fall through a warm layer of air before they hit the earth, they melt to produce rain. If they don't, they fall as freezing rain, ice pellets or snow.

In clouds where there are no ice crystals, like those in the hot tropics, water droplets keep splatting together and sticking. Each drop gets bigger and heavier until . . . it starts to fall. Small raindrops can fall as fast as 9 m (yards) per second. That's about as fast as you can bicycle at top speed!

RAINY-DAY BLUES . . .
AND REDS AND GREENS

Do you feel blue when it rains? What about red or green? While there's never been a case of blue rain, there have been many other coloured rains:

- Red rain (and snow) fell in Italy in 1755, coloured by small bits of reddish soil in the air.
- Black rain fell in New York State and brown rain fell in Vermont after a great dust storm filled the air with swirling dust in 1933.
- As recently as May 5, 1987, green rain fell on Moscow. It was probably coloured by green pollen in the air.

These coloured rains happen when water droplets pick up bits of coloured dust or pollen as they fall.

TAKE A CLOSE LOOK AT A RAINDROP

It's difficult to catch a raindrop and measure it, but here's a way to preserve one so that you can take a closer look.

You'll need:

o *a shallow lid from a box (a shoe-box lid works well)*
o *flour*
o *a knife*
o *a fine mesh sieve*
o *a bowl*

1. Pour flour into the lid to a depth of about 1 cm (1/2 inch).

2. Use the knife to flatten the top of the flour so that it's level.

3. Put the box out in the rain until some raindrops have splattered into the flour.

4. Back in the house, pour the flour through the sieve. (Put the sieve over a bowl to catch the flour so you can use it for cooking later.) See any lumps in your sieve? They're preserved raindrops.

How big are the drops you caught? Chances are you'll have several sizes. Scientists have discovered that some raindrops are five times bigger than other ones.

small (2 mm / 1/13 inch)
medium (3 mm / 1/8 inch)
large (6 mm / 1/4 inch)
extra large (1 cm / 1/3 inch)

What shape are your raindrops? If you could observe a raindrop before it hit the ground, you'd be in for a surprise. Raindrops aren't shaped like tear-drops. They're actually shaped like miniature hamburgers. Why? Although raindrops are round when they first start to fall, wind resistance on the way down flattens them slightly into burgers.

HOW WET CAN IT GET?

If you lived on Mount Waialeale in Hawaii, you'd have lots of opportunity to study raindrops. Gazillions of them fall ever year. In fact, Mount Waialeale is the wettest place on earth, with an average annual rainfall of 11 684 mm (38 feet). If all that rain fell at once, it would submerge a three-storey building. A hundred years ago Cherrapunji, India, got more than twice that much rain (26 461.2 mm/87 feet) in one year.

Scientists keep track of how much rain falls by using a rain gauge. Here's how you can make a simple rain gauge.

You'll need:

o *a glass with straight, rather than curving sides*
o *a ruler*
o *clear sticky tape*

1. Hold the glass at eye level. Position the ruler as shown. (You want to measure the contents of the glass, not the bottom of the glass.) Tape the ruler in place.
2. Put your rain gauge out in the rain on a flat surface.
3. After a rainfall, record how much rain fell. Check the local paper the next day to find out the official rainfall. How accurate was your rain gauge?

DON'T FORGET YOUR UMBRELLA!

You set out for school on a cloudy day. Will it rain or won't it?

You've got too much to carry, so you decide to leave your umbrella in the closet. Two blocks from school it starts to rain. You hold a notebook over your head in an attempt to stay dry. All you get is wet homework. You start to sprint. Your shoes squish. By the time you steam into the school yard your hair and clothes are ruined—and so is your mood.

What you need (besides an umbrella) is a way to predict rain before it starts to fall. Over the years people have come up with some rain predictors that they swear by. How well do you think they work?

Your built-in rain predictor

You have a rain predictor growing on the top of your head. As the air becomes more and more humid (often a sign of rain to come) your hair lengthens by as much as 2 per cent.

In 1783 a hair hygrometer was invented to predict rain. First a long hair was boiled in chemicals to get rid of its natural oils. Experiments showed that blond hairs worked best. Then the hair was tied to a needle. The needle moved as the hair lengthened and shortened, indicating when to expect rain.

Plant predictors

British gardeners watch the scarlet pimpernel plant to find out whether it will rain. If the relative humidity (amount of water in the air) climbs to 80 per cent, the pimpernel flower closes up, probably to shield its inner parts from water damage. Then British gardeners get out their umbrellas.

Other plants may also predict rain. See for yourself how accurate this plant prediction is:
• red and silver maples and poplar trees turn up their leaves when rain threatens.

Kill a spider and it'll rain

Nothing you can do to a spider (or any other animal) will bring on rain. But you *can* find out when rain is coming by watching animals, especially insects. Bees stick close to home when it's about to rain. How do they know rain is coming?

Scientists believe that the bee's sensitive antennae can detect increases in humidity, which often mean rain. When bees know rain is coming, they stay home.

Ants and flies are also thought to be good rain predictors. Next time the weather forecasters call for rain, check out two old beliefs:
• Expect stormy weather when ants travel in lines and fair weather when they scatter.
• Slap at a fly that lands on your stockings. If it flies away and comes back again, it'll rain.

Turn on your radio

If your AM radio spits out static squeals and crackles, don't just expect rain; get ready for a thunderstorm. The radio is picking up electrical discharges produced by nearby lightning. You may not be able to see the lightning; it may be zapping from cloud to cloud rather than down to the ground. But be prepared. Crackles on the radio (scientists call these "sferics") are often followed by a *KABOOM* on the ground.

PICNICS OR PUDDLES?

Today, sophisticated instruments predict the weather. But it's still difficult to forecast rain accurately. You may have heard weather forecasters talking about the "probability" of rain. A 10-per-cent probability means there is a one in ten chance of rain falling during daylight hours where you are. A 90-per-cent probability of rain means . . . cancel the picnic.

MAKE A RAINBOW IN YOUR BACKYARD

If someone said, "Ratting on your goofy brother is vulgar," what would they be talking about? Minding your manners? Maybe. But they might also be talking about rainbows.

The first letters in "Ratting on your goofy brother is vulgar" spell ROYGBIV. That's the order of the colours in a rainbow: (R)ed, (O)range, (Y)ellow, (G)reen, (B)lue, (I)ndigo and (V)iolet. The colours in a rainbow always appear in this order. Now you have an easy way to remember them.

People have always been fascinated by rainbows. These glowing "sky bridges" seemed so amazing that stories were made up to explain them. One of the most famous was the legend of the pot of gold at the end of the rainbow. You can find out if there's any truth to this old story by making a rainbow in your backyard.

All you need is a hose with a sprinkler nozzle. Stand with your back to the sun and send a spray of fine mist up in an arch in front of you. For best results, try this late in the afternoon when the sun is low in the sky. Hold the hose about eye level. With a little experimenting, you'll see a rainbow. Don't forget "ratting on your goofy brother is vulgar."

What makes the colours? Sunlight contains all the colours you see in a rainbow, but usually they're all mushed together so you can't distinguish them. When sunlight enters a raindrop or spray from a hose, the light rays get bent and the colours split apart so that you can see each one.

A good time for seeing rainbows is early in the day or late in the afternoon, when the sun is low above the horizon. The clearest rainbows happen on days when the raindrops are large. If you're really lucky, you might even see a double rainbow. The second rainbow is located outside the main rainbow. Its colours are paler and their order is backwards.

If you don't have any luck spotting a rainbow, try hunting down a "moonbow." Look for a moonbow after it rains on a full-moon night just after the moon has risen. That will give you the ingredients you need for a moonbow or a rainbow: water droplets in the air and light. Can you think of another time when you have those two ingredients? (Answer on page 93.)

SUN SEPARATOR

Want to break some sunlight into the colours of the rain-bow? Try this.

You'll need:

○ *a pocket mirror*
○ *a white wall (or a piece of white paper)*
○ *a shallow pan of water*
○ *sunshine*

1. Put the pan of water in a sunlit place opposite a white wall.

2. Hold the mirror at one end of the pan so that the sun-light strikes it.

3. Move the mirror until you see the colours of the rainbow on the wall.

BRRRRRRRRR!

If you live in an area where the winters are cold, here are some of the objects you might find falling on your head, freezing your fingers and sticking to your boots. To make the best of winter, try tobogganing, skiing or making snowmen out of them.

Blechhhh!

You're out for a walk in the winter when all of a sudden you get hit by something cold and wet. It's not quite rain, but it's not snow either. What is it? Ice pellets, or sleet as it's called in the United States. Ice pellets form when raindrops freeze as they pass through a layer of cold air before they hit the ground. They also form when snowflakes melt and refreeze on their way down. Pull up your collar and watch your step!

Let it snow

During the winter of 1921, the people of Silver Lake, Colorado, were treated to a real winter wonderland. In just 24 hours, 193 cm (76 inches) of snow fell on their town.

When it snows, you have two choices: you can wait for spring or you can shovel it. The weight of 20 cm (8 inches) of snow on a sidewalk two car lengths long is about 200 kg (440 pounds). During Silver Lake's record snowfall, the snow would have weighed almost 2 tonnes (2 1/5 tons). That's more than what two large polar bears weigh!

LEARNING FROM THE ANIMALS

How do animals survive winter without warm homes and clothes? Here are some of their survival tips:

- conserve energy. Bears and other animals hibernate during winter. Even animals that don't hibernate are less active when it gets cold. Why? Because that way they use less energy. The energy they save can be used to keep them warm.

- insulate yourself. Animals put on an extra layer of fat and grow thicker fur or extra feathers. You can insulate yourself from the cold by wearing layers of clothes. Air that is trapped between the layers warms up and keeps you warm.

- snuggle up. Take a lesson from chickadees and garter snakes. They cuddle up in large groups to minimize heat loss—but not together.

Icing

When rain falls and then freezes as it hits objects on the ground, you get—trouble! Freezing rain can transform your town into one big skating rink. People slip and fall, cars slide and spin their tires, and trees groan under the weight of the ice. Heavy sheets of ice sliding off buildings can dent cars and break windows. Even a thin coating is enough to snap telephone and power lines. The result: a long cold wait until power is restored.

CATCH A FALLING SNOWFLAKE

How can you catch something that melts when you touch it? It's not as difficult as it sounds.

You'll need:

o *a clean microscope slide or glass camera slide*
o *spray-on clear plastic or lacquer (you can get this at a craft or hardware store)*
o *tweezers*
o *magnifying lens*

1. Put the slide and spray into the freezer and let them cool completely, overnight if possible.
2. When you want to catch some snow, remove the slide from the freezer with the tweezers and spray a thin coating of plastic spray on one side. Tilt the slide to get rid of extra plastic. Always handle the slide with tweezers, otherwise the warmth of your hands will heat up the slide.
3. Put the slide outside and wait till a snowflake falls on it.
4. Using the tweezers, put the slide in a protected place so that no more snowflakes will fall on it. Leave it outside for one hour.
5. After an hour, bring the slide inside and use a magnifying glass to look at the plastic impression of your snowflake.

40

Does your snowflake have six points, like the ones you can cut out of folded paper? Match your snowflake with one of the seven basic shapes. Your snowflake probably won't look exactly like one of these, but there will be a family resemblance.

PLATES

STELLARS

IRREGULAR CRYSTALS

COLUMNS

SPATIAL DENDRITES

NEEDLES

SNOW WHAT!

What do you know about snow? Test your snowledge with this quiz. Answers on page 93.

1. Pink snow has fallen on parts of Western Canada.
   a. True
   b. False
2. A large snowflake falls at a speed of:
   a. 0.5 km/h (1/3 mph)
   b. 5 km/h (3 mph)
3. Sometimes it's too cold to snow.
   a. True
   b. False
4. The largest snowflake ever recorded was:
   a. The size of a hockey puck
   b. The size of a bread and butter plate
   c. Bigger than a long-playing record

5. If you melted down a pail of water containing 25 cm (10 inches) of snow, how much water would you be left with?
   a. 12 cm (5 inches)
   b. 2.5 cm (1 inch)

IT'S IN THE WIND

You can't see it but you can see what it does. It ruffles your hair, tips over garbage cans and screeches through bare branches like a ghost in a bad mood. It flutters, blusters and howls all around us day in and day out. What makes the wind? What's floating around in it? Hold on to your hat and read on.

WHERE DOES THE WIND COME FROM?

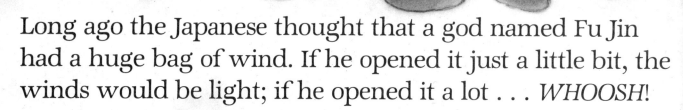

Long ago the Japanese thought that a god named Fu Jin had a huge bag of wind. If he opened it just a little bit, the winds would be light; if he opened it a lot . . . *WHOOSH!*

Other peoples thought the winds were kept in a cave or created by a god who squeezed his bellows in the sky. Today we know better.

Wind is created when air flows from an area of high pressure to an area of lower pressure. When you press the button on an aerosol spray can, air that has been packed in the can under high pressure is released and blasts out into the surrounding air, which has a lower pressure. Presto!—you've created a miniature wind.

There's no big aerosol can in the sky. But there *are* areas of high and low pressure. The sun heats up some parts of the earth more than others. The air above these "hot spots" is warmed, too, and rises. This rising air forms a low-pressure area. Air flows from higher-pressure areas to these low-pressure areas. Unless it gets bumped off course.

What's big enough to change the direction of the wind? The earth's spinning, for one thing. It forces the winds to turn right in the Northern hemisphere and left in the Southern hemisphere. Mountains, buildings and forests can alter the direction of the wind too, just as rocks in a stream change the water's flow.

FIND OUT WHERE THE WIND COMES FROM

North winds blow from the north, south winds from the south and so on. Winds get their names according to where they come from. You can find out which wind is blowing on you with this simple-to-make wind vane.

You'll need:

o *a ruler*
o *a felt pen*
o *a square piece of wood*
o *a hammer*
o *2 nails 7.5 cm (3 inches) long*
o *a cork*
o *a cap from a pen*
o *a feather with a long shaft*
o *a compass*

1. Measure the halfway point on the four edges of your block of wood. Use the felt pen to mark the directions North, South, East and West at the halfway points as shown.

2. Draw two diagonal lines to find the midpoint of your block of wood.

3. Hammer one nail part way in at the midpoint. Make sure the nail doesn't get bent. It should stick out about 5 cm (2 inches).

4. Use the other nail to dig a hole in the cork large enough for most of the pen cap to fit in. If it doesn't fit snugly, glue it in place.

5. Glue the feather across the top of the cork as shown. Make sure the feather is in the middle of the cork.

6. Put the cork, feather and pen cap over the nail that's stuck into the wood block as shown.

7. Use a compass to line up North on your wind vane with magnetic North.

8. Watch your wind vane as the wind blows. The pointed quill end of the feather will always point into the wind and tell you what direction the wind is coming from.

WINDS ON THE MOVE

If you walk outside right now, there's a good chance you'll be hit in the face by wind that has come from hundreds of kilometres (miles) away. Winds are never still. If they were, they wouldn't be winds—they'd be calm air.

Although you can't feel the difference, there are hundreds of different winds. Some, like the ones you see on this earth map, blow thousands of kilometres (miles) year round. These are the big winds on earth.

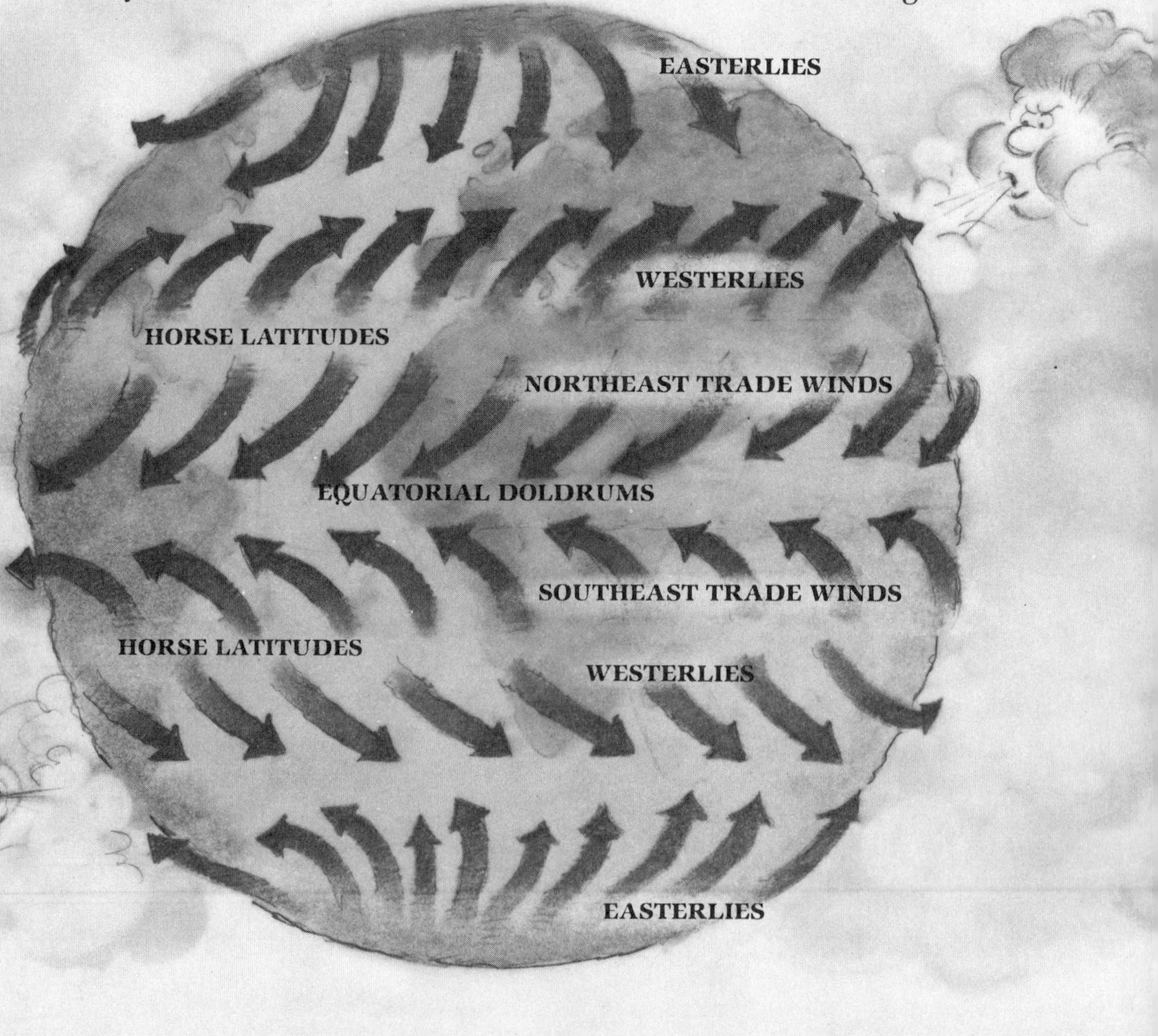

EASTERLIES

WESTERLIES

HORSE LATITUDES

NORTHEAST TRADE WINDS

EQUATORIAL DOLDRUMS

SOUTHEAST TRADE WINDS

HORSE LATITUDES

WESTERLIES

EASTERLIES

Other winds blow only at certain times of the year and over only a country or two. In India, for example, wet monsoon winds sweep in from the Indian Ocean during the summer and dump their moisture in torrential rains. In the winter, the winds reverse directions and blow hot and dry. Monsoon winds give India a wet season and a dry season.

Finally, there are local winds that last for a few weeks — or just a few hours. If you live near the sea or a large lake, you've probably felt a breeze coming from the direction of the water during the day. This breeze starts up because the air above the land is heated more than the air above the water. The heated air above the land has lower pressure than the cooler air over the water. This cooler sea air from higher-pressure areas flows towards the lower-pressure areas over land. At night, the process reverses. The breeze flows from the land towards the water.

Winds that always happen at the same time or in the same way have been given names. Some wind names sound as if they belong in a zoo: elephanta, haboob, zonda and williwaw, for example. Others, such as xlokk, simoom, kwat and bhoot, sound more like Martian fast food. (To see where these winds blow, turn to page 93.) If you had a chance to name a wind, what would you call it?

But how fast are they moving?

You don't need fancy instruments to know how fast the wind is moving. You just need to know the Beaufort scale. It's a wind-rating system named after its inventor, Sir Francis Beaufort, a British admiral. It describes how the wind behaves at various speeds. A calm day rates a 0, while a hurricane rates a 12. (See page 91 for the whole Beaufort scale.) So, for example, if you hear the wind whistling through telephone wires and have a tough time taming your umbrella, you know you're being buffeted by a wind rated at force 6, which can travel up to 50 km/h (31 mph).

On April 12, 1934, the winds at the peak of Mount Washington in New Hampshire blasted right off the top of the Beaufort scale. They set a record, reaching 371 km/h (231 mph).

CAUTION: WINDS AT WORK

If you could harness the winds the way you harness a horse, you could get them to do a lot of work for you. Why? Moving air has energy. Want to see some wind power for yourself?

You'll need:

o *a piece of thick paper*
   15 cm (6 inches) square
o *a pencil*
o *a ruler*
o *scissors*
o *tape*
o *a thumbtack*
o *a popsicle stick or*
   a long cardboard tube

1. Draw diagonals on your paper as shown and cut as far as the dots.
2. Bend in the corners of the paper one by one and tape them in place.

3. Put the tack through the middle of the paper and into the stick or tube. Don't tack it too tightly or your whirligig won't whirl.
4. Point the whirligig into the wind and give it a spin to get it turning. Then watch the wind energy take over.

If you made a giant whirligig out of sturdy materials, you could use the turning motion to turn other things or even make electricity. That's what windmills do. If the world's wind power were harnessed, there'd be enough energy to meet all the world's needs.

OTHER WINDY JOBS

Sure, winds cool us off and sometimes even warm us up. And they blow away clouds and fogs. But winds also do other things.

Wipe that face off your face

Wind wears away the things that it blows over —including this sphinx's face. This wearing away is called erosion. Wind erosion can blow good soil off farmers' fields and deposit it hundreds of kilometres (miles) away. In the spring of 1989, so much topsoil was blown off fields in parts of Saskatchewan that it will take 500 years for nature to replace the soil.

DANDELION SEED

MAPLE KEY

Hitch-hikers

Winds sometimes whisk seeds to new growing spots. Here are two champion seed hitch-hikers. One spins like a helicopter and the other floats like a parachute. Which is which?

Sailing, Sailing

Once the only way to cross the Atlantic was by sailing ship. If the winds didn't blow, the ships didn't go. Sailors gave the various winds names that suited them. The main east-west winds were called the trade winds. They were so reliable that merchants could count on them to deliver their shipments on time. Zones without wind had to be avoided. One windless zone was named the horse latitudes. If ships carrying horses got stuck there and they ran out of horse feed, the sailors would have to throw the poor beasts overboard.

Today, most people cross the ocean by airplane. Eastbound planes can use jet streams much as sailors use the winds. Air in the jet streams can travel as fast as 360 km/h (224 mph) and can shorten a flight from North America to Europe by an hour.

IT'S IN THE AIR

There's more to air than meets the eye. To see what the air contains, try this.

You'll need:

○ *a clean shallow pan*
○ *a coffee filter*
○ *a coffee-filter holder*

1. Fill the pan with water and put it outside in the open. Leave it for two days.

2. Put the coffee filter in the holder and pour the pan water through it. What's caught in the coffee filter? Nothing? You're lucky. In most towns and cities the filter — and the air — will be filled with particles. The specks of stuff you can see are the giant ones. There are many more invisible particles. They are so small that you could jam-pack billions of them into a thimble. In a big city there may be up to 100 000 visible and invisible particles in a thimbleful of air. If you want to go some place cleaner, try the middle of an ocean or the North or South Poles. There, you'll usually only find about 300 particles in a thimble of air.

Where does all this stuff come from? Ashes and soot are belched into the air by volcanoes and forest fires. Pollen, bits of spider web, tumbleweed and other plant bits float upward on air currents. And tiny pieces of rock and salt from the ocean water are whooshed into the air by the wind. This natural pollution is necessary. It helps clouds form and rain fall.

The problem is, we humans have added more pollution by driving cars, heating buildings and operating machinery and factories. This pollution is dangerous. It mixes with water in the sky and falls as acid rain. It rips holes in the protective ozone "sunscreen" that surrounds the earth. And it changes our weather.

You've probably caught some of the pollution that causes all this damage in your coffee filter. To find out more about what it's doing, turn the page.

SNOW, WHAT'S IN IT?

You can test the amount of pollution in the snow around your home much the same way you tested the rain. Scoop cupfuls of snow from various places in your yard. Take one sample from near a road. Let the snow melt and pour each sample through a separate coffee filter. How much dirt and pollution did you trap?

51

S.O.S. PLANET EARTH

Poor old planet Earth! Pollution is making it sick. And when Earth gets sick, so does its weather. Here are the three main "pollution diseases" attacking Earth's weather —and what you can do to help stop them.

1. THE NO-OZONE SUNBURN

The earth is a bit like you. Too much sun can result in sunburn. Up until recently the earth has been protected from the sun's harmful rays by a layer of ozone gas that acted like sunscreen lotion.

Then pollution started attacking the ozone layer and ripping holes in it. The first hole was spotted over the Antarctic in 1982. It was as big as the United States. Four years later, a smaller hole was found over the Arctic. These holes and thinning elsewhere in the ozone layer let in harmful sun rays that can cause skin cancer in humans, damage sea creatures that live near the water's surface and reduce harvests of important food crops such as wheat, rice, corn and soya beans. That's some sunburn.

What's eating the ozone layer?

Chlorofluorocarbons, or compounds that contain chlorine, fluorine and carbon. Until recently these were used in aerosol spray cans, but now the governments of the United States and Canada have banned them. But chlorofluorocarbons are still used — in refrigerators and air conditioners and in the foam packaging fast food often comes in.

You can help protect the ozone layer by going on strike against foam packaging. Let grocery stores and restaurants know you don't like it and why. Ask them if there isn't something else they could use.

At home, try recycling foam packages. Use burger boxes as mini-coolers to keep school lunch foods chilled or store stuff in them. Grow seeds in the foam egg cartons you have now. (Next time buy eggs in cardboard cartons.) You can even make a great glider out of a leftover Styrofoam meat tray.

THE F(LYING) M(EAT) T(RAY)

You'll need:

o *a clean Styrofoam meat tray*
o *a felt pen*
o *sticky tape*
o *scissors*
o *a ruler*

1. Trim off the curved edges of the meat tray and cut the flat piece that remains into a square.

2. With a marker, draw a triangle on the Styrofoam. Cut along the lines with scissors. You will use the big triangle as the FMT's wings and one of the small triangles as the rudder. Throw away the other small triangle.

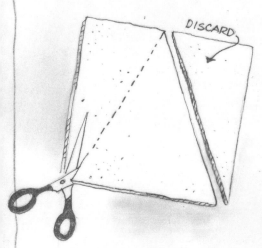

3. Cut a slot to the midpoint of the large triangle. The slot should be just wide enough so that another piece of Styrofoam will fit into it snugly.

4. Cut a slot not quite to the midpoint of the rudder as shown. This slot should be the same width as the other one.

5. Cut the front of the rudder off at an angle.

6. Push the rudder into the wing so that the two slots fit together.

7. If the rudder is loose, tape it to the wing. When fitted in, the rudder should stick out a little past the end of the wing. Make a fold line on the rudder as shown.

8. Make a 1 cm (½ inch) long cut on each side of the rudder. Then make a fold line as shown. These flaps are your elevators. Raise or lower them to get your FMT flying straight.

53

2. ACID INDIGESTION

Clouds and rain are the weather's digestive system. Water is drunk from lakes, rivers and oceans and peed back out as rain. What happens when acid poisons the system? Acid rain pours down, killing lakes and the creatures that live in them, attacking many crops and even eating away buildings, bridges and statues.

The acid in the rain comes from chemicals in car exhausts and in the smoke from burning coal and oil. Since these chemical particles are very small, they can travel long distances on the wind before they fall back to earth. This means that acid pollution created in one place or country may fall as acid rain in another. For example, half of the acid rain that falls in Canada is produced in the United States. Acid rain from Canada ends up in Scandinavia.

There are ways to reduce acid rain. Industries that use coal or oil can clean up their act. They can put scrubbers on chimneys to filter out chemicals in smoke that cause acid rain. For how you can help, see the box on next page.

3. GREENHOUSE FEVER

What happens when you have a fever? Your temperature climbs and you feel H-O-T, right? You could say the Earth is coming down with a fever. Little by little the Earth's temperature is rising. Earth is becoming hotter. What's causing this fever?

If you ever walk through a greenhouse, you'll notice that it's hotter in the greenhouse than outside. That's because heat from the sun is trapped inside by the glass.

The same thing is happening up in the sky. Only instead of glass, there are gases that trap the sun's heat. These "greenhouse gases" are a form of pollution that happens when we burn coal, oil and other fuels. Other gases, such as the chlorofluorocarbons, add to the problem.

At first it might sound great to have warmer weather. You could put away your boots and heavy jackets and spend more time at the beach. But weather that's even a few degrees hotter can:

- turn good crop-growing areas into near deserts
- wipe out wildlife as people turn animals' wilderness homes into food-growing areas
- cause water shortages in some areas and too much rain in others
- wipe out forests in some areas before they have a chance to spread into new growing areas
- melt the cap of ice surrounding the North and South Poles and add water to the oceans, causing flooding in low-lying coastal areas.

Living in a greenhouse might be okay for a tomato, but it's a big problem for planet Earth. To find out how you can help stop this "greenhouse effect," see the box on this page.

HELP CURE THE EARTH'S ACID INDIGESTION AND FEVER

There's no miracle cure for pollution diseases like acid rain and the greenhouse effect. But there are things that you can do:

- Write letters to the President or your legislators asking for laws to cut the pollution that causes acid rain and the greenhouse effect. Since these are global problems, write to other world leaders, too. Every time officials get a letter they assume that 20 other people feel the same way. These people need our votes to keep their jobs, so letters are powerful weapons.
- Turn your family car into a museum piece. Let it sit in the garage gathering cobwebs while you walk or pedal or take the bus. Remember: car exhaust contributes to acid rain and the greenhouse effect.
- Buy food, not packaging. Who needs all that foam packaging anyway?
- Don't throw it away—recycle it. Let's say you take a plastic shopping bag back to the store for your next load of groceries. That means a new shopping bag doesn't have to be manufactured, and fewer bags will end up in the garbage.

KILLER SMOG

If you live in a big city, you've breathed smog. Londoner Harold De Veaux named smog in 1905 by combining "**sm**oke" and "**fog**," the two main ingredients in smog back then. Today's smog is slightly different; it's made up of chemicals "cooked" by the sun.

Most smogs are unpleasant but harmless. Some are dangerous for people who have breathing problems. But in December 1952, "Killer smog" blanketed London, England. Five days later, when the smog lifted, 4000 people were dead.

Here is one young Londoner's smog diary.

Thursday, December 4: London smells terrible. Must be the gases from the factories. Noticed some fog today . . . Annie says her eyes are hurting.

Friday, December 5: The air is thick and yellow. Grannie called to say her asthma is bad. Annie's eyes are still stinging.

Saturday, December 6: We tried to get to Grannie's this afternoon but couldn't make it. The smog is so brown and thick now that the cars and buses can't see to move.

Sunday, December 7: Dad is working day and night at the hospital. Lots of sick people because of the smog.

Mum wouldn't let Annie go outside, but it's not much better inside. The smog oozes in and leaves a sticky black coating over everything. . . . Grannie needs a doctor but no one can get to her. George Drew went to the movies and said they sent everyone home. The smog in the movie house was so thick no one could see the screen.

Monday, December 8: A doctor finally got to Grannie and she's all right. He couldn't see for smog. So he asked a blind patient to guide him. The blind man used his cane. What a hero that man is. Another good thing happened. School was closed because of smog. Hurray!

Tuesday, December 9: The smog is blowing away. . . . The newspaper said thousands of people died. Thank goodness that doctor made it to Grannie.

SMOG IN A JAR

The foggy stuff that you'll make in this experiment is a mild version of the killer smog that covered London in 1952.

You'll need:

○ *a jar*
○ *water*
○ *aluminum foil*
○ *ice cubes*
○ *a spoonful of salt*
○ *a strip of paper about the length and width of a pencil folded in half and twisted*
○ *a match*

1. Pour water into the jar and slosh it around. Dump the water out.
2. Tear off a piece of aluminum foil big enough to fit over the mouth of the jar and shape it to fit the jar's mouth.
3. Remove the foil and put ice cubes on it, in the centre, where the mouth of the jar goes. Carefully sprinkle the salt over the ice cubes.
3. Ask an adult to help you light the piece of paper with the match and drop it into the jar. Quickly put the foil with the ice cubes and salt over the mouth of the jar and seal the foil tightly. Look but don't breathe it in.

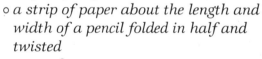

WEIRD WEATHER

What can pluck a chicken, rip the bark off a tree, drive a straw into a piece of wood and send cars flying through the air? A tornado. Over the years tornadoes have done everything from digging up a farmer's potato crop to planting a cow upside-down in a field. But tornadoes aren't the only weird weather around. In this section you'll find out about tropical killers called hurricanes, giant hailstones, doughnut-shaped snow rolls and more.

COLLECTING TORNADOES

Some people collect baseball cards or stamps or bottle caps. But not Bonnie and Bob. They collect tornadoes. These huge wind funnels develop without warning and are strong enough to suck cars and even houses off the ground.

Tornadoes, or "twisters" as they're sometimes called, zigzag across the country like whirling dervishes, wrecking everything in their path. How do you collect a tornado? What do you do with it once you've got it? Ask Bonnie and Bob.

How in the heck did you get started collecting tornadoes?

Bonnie: 'Bout 10 years ago we saw a tornado rip through a town. I tell ya, it was somethin' else. It started with a big black cloud. Then a whirling funnel reached down from it. It kinda looked like an elephant's trunk. That twisting trunk ripped roofs off houses and flipped cars into basements. We even saw straws from a broom stuck right into a telephone pole. We wanted to get a closer look, so we started drivin' around lookin' for tornadoes.

Did you spot one?

Bob: Me and Bonnie drove all over Tornado Alley—that's a chunk of country from northern Texas, Oklahoma, Kansas and right up into southern Iowa. But we didn't see a single tornado. Then we met this old guy who had been chasin' tornadoes for 20 years. He sat us down and gave us the dirt on tornadoes. Seems you only get 'em when cool dry air moves in on hot humid weather. The best time to look is between three in the afternoon and dinner time. Nowadays, we spot six or so a year.

Right up close?

Bonnie: As close as we wanna be, let me tell you. One day in Oklahoma we almost got *too* close. We saw a tornado just startin' up. Danged if it didn't head right for us. Whooooee! I tell ya it was somethin' else seeing that black funnel coming our way. We could see mud, chunks of wood and papers and even a chair whippin' around at the bottom of the funnel. It must have touched down and picked up that stuff. And was it roarin'! It sounded like a million bees all buzzin' together. It was pourin' down rain as we headed out of there — fast. Most tornadoes travel about 50 to 70 km/h (30 to 45 mph). But this one was faster 'n a car on the highway. We were lucky when it changed its mind and gave up on the chase. We found out later that it was an F3, with winds of about 200 km/h (125 mph).

An F3?

Bonnie: That's a bad one. Tornadoes are rated — bit like earthquakes — from F0 to F5. A F5 doesn't sound so bad, till ya consider it's strong enough to lift a wood house into the air or send a car flying the length of a football field.

What do you do with a tornado once you've spotted it?

Bob: Why, we collect it . . . on film. Then we send our photos in to the weather office, so they can get more information on tornadoes. Learnin' about tornadoes won't stop 'em, but it might just help us figure out when they'll start up so we can get out of their way.

Remember Toto?

"Tornado collectors" like Bob and Bonnie give meteorologists much-needed information about tornadoes. But recently weather watchers have been getting help from a box named TOTO. It stands for *To*table *To*rnado *O*bservatory, and it's filled with weather-measuring equipment.

If you've seen the movie *The Wizard of Oz*, you'll know that a dog named Toto and its owner, Dorothy, were sucked up by a tornado. Scientists hope that the new TOTO will get hit by a tornado, too. But instead of being swirled away to the land of Oz, it'll stay put and record wind speed, pressure and other information as the tornado passes over it.

HURRICANE!

What do Agnes, Hazel and Gilbert have in common? They're huge tropical storms called hurricanes. All three have whipped cities and towns with roaring winds and flooded them with torrential rains.

Hurricanes start out as ordinary storms over the warm tropical oceans. They grow bigger and bigger as they absorb heat and moisture from the warm ocean water. Differences in air pressure start winds spiralling, creating a doughnut-shaped storm system that can be as wide as 600 km (375 miles).

Scientists watch hurricanes carefully. If a hurricane starts heading towards land, they want to warn the people in its path. They know that hurricanes travel an average of 20 km/h (12 mph), so they can tell people how much time they have till the hurricane hits.

Waiting for a hurricane to arrive is a bit like waiting for a school-yard bully. You know what you're in for, but you don't know when you'll get it. Like the approaching bully, a hurricane may charge straight ahead or veer off in another direction at the last minute. So, for the nine days or so that a hurricane lasts, people in the area wait and watch.

If a hurricane hits land, the first thing you experience are the weak edges of the storm. But as the doughnut-shaped storm moves overhead, the winds travel so fast they could break highway speed limits. They uproot trees, rip off roofs and whip up giant waves that cause flooding. As much as 25 cm (10 inches) of rain can pour down in just one day, enough to fill a pail.

Strangely, in the middle of this chaos, there is a calm. Like a doughnut, a hurricane has

Hurricane, cyclone or typhoon?

Hurricanes are the same thing as cyclones and typhoons, except that cyclones occur in the Indian Ocean and typhoons in the China Sea. The word hurricane comes from Hunraken, the storm god of the Mayas of Central America. Typhoon comes from *ty fung*, meaning "great wind."

nothing in its centre—no winds, no rain—just calm air. In this "eye" of the hurricane the winds may be calm and the sky blue. The eye sometimes serves as a huge bird cage, trapping birds who travel with it to keep from being sucked into the storm. The calm of the eye has been known to lure people out of their homes. They think the storm is over, and then *pow*! They are hit by the other half of the dough-nut's ring.

Over the years, hurricanes have killed millions of people, destroyed whole communities, flattened crops and sunk ships. About six hurricanes hit in the Atlantic Ocean every year. To tell the world's hurricanes apart, meteorologists give them names.

Don't expect to have a hurricane named after you if your name is David or Allen or Hazel. Famous hurricanes have already been given those names. To avoid confusion, meteorologists agreed not to use the same names again. The names have been "retired" just as hockey sweater number four was retired by the Boston Bruins when Bobby Orr stopped playing hockey.

The idea of naming hurricanes started in the United States in 1953. At first, all hurricanes were named after girls, but in 1979 boys' names were added. Now, if the first hurricane of the year is named after a girl, the next would be named after a boy and so on. The next year, boys lead the list.

The names follow the letters of the alphabet. In 1990, for example, meteorologists decided that the hurricanes for that year would be named: Arthur, Bertha, César, Diana, Edouard, Fran, Gustav, Hortense, Isidore, Josephine, Klaus, Lili, Marco, Nana, Omar, Paloma, René, Sally, Teddy, Vicky and Wilfred. There are no names starting with the letters Q, U, X, Y or Z because there aren't a lot of names beginning with those letters. So far meteorologists have never run out of names before the end of a year.

GREAT BALLS OF ICE

How big was the world's largest hailstone? Was it as large as a golf ball, a tennis ball or a grapefruit? If you guessed a grapefruit, you'd be close. A hailstone bigger than a grapefruit smashed into the town of Potter, Nebraska, on July 6, 1928. Other places have been hit with hailstones the size of golf or tennis balls.

How do such huge chunks of ice form high up in the sky? Hailstones start out the same as many raindrops—as tiny crystals of ice in the upper part of clouds. Water droplets then stick to these crystals, and as they get larger and heavier, they start to fall through the cloud.

Sometimes hailstones fall straight down to the ground. But other times, they get trapped in the cloud and are whooshed up and down by strong air currents. Each time they travel up and down in the cloud, another layer of water sticks to the hailstones and freezes. Eventually the hailstones grow so heavy with layers of ice that they drop to the ground. Or, if the updrafts are strong enough, they come shooting out the top of a cloud like peas out of a pea-shooter!

HUMAN HAILSTONES

The updrafts that carry the hailstones back up through the cloud are strong enough to lift you up. In fact, they once lifted five men! One day in 1930, five German glider pilots found themselves gliding through a thunderstorm. Fearing their plane might be ripped apart by the storm, they tried to parachute to safety. But updrafts in the storm carried them up like human hailstones. Down they fell and up they wooshed getting covered with layer upon layer of ice. All five finally fell to earth, but only one lived to tell the story.

Not all hailstones are as round as grapefruits. Some are conical, like carrots; others may look more like mushrooms or slices of cucumbers. But all hailstones have something in common with onions. They are made up of layers.

You can make a hailstone in your freezer and then cut it apart to see the layers.

You'll need:

o *a small, tightly packed snowball about the size of a golf ball (wet snow works best)*
o *cold water*
o *a plant sprayer*
o *a knife or saw*

1. Spray the snowball with water and put it in the freezer.
2. When the water is frozen, spray the "hailstone" again.
3. Keep spraying and freezing your hailstone until it has a thick layer of ice. Repeat these steps for several days.
4. Ask an adult to help you saw through the hailstone and look at the layers.

IT CAME FROM OUTER SPACE

Suppose aliens from a distant planet landed on Earth and brought some of their weather with them. They might have weird weather like the phenomena you'll read about on this page. But wait a minute . . . are the six weather weirdos on this page science fiction or are they fact? Were they brought by aliens from outer space, or do they happen right here on Earth? Try your luck at guessing, then turn to page 93 for the answers.

1. Dry Rain

A dark cloud forms. Looks like rain. But instead of rain, virga falls. What's virga? Rain that starts to fall but evaporates as it leaves the cloud. A cloud that "rains" virga looks a bit like a stuffed toy with some of its stuffing hanging out the bottom. Is virga for real, or does it belong on Venus?

2. Lightning balls

Imagine lightning that comes not in bolts but in balls. Ball lightning doesn't just zap down from clouds. A glowing ball about the size of a beach ball or bigger zigzags from place to place, even inside houses, as if it has a mind of its own. Is ball lightning a "Star Trek" invention or is it fact?

3. Snow rollers

Does snow always stay put when it falls? No way. Snow rollers are thick doughnuts of snow that roll along the ground on their sides, like empty tin cans. What causes them? The wind whips up a layer of snow that curls under, picking up more snow as it rolls. Are snow rollers a circus trick on Saturn, or do they roll around here on Earth?

4. Hit by a herbie

A herbie is like a wall made of snow that hits without warning. One minute you're skating in the winter sunshine, and the next minute you're blinded by a mini-blizzard so ferocious you can barely see your mittens at the end of your arms. Do herbies hit Martians — or humans?

5. Glow-in-the-dark clouds

You'd have to be cloud watching at night to see one of these. It's a bluish-silver cloud that glows in the dark and gets brighter as the night goes on. It's shaped a bit like a cirrus cloud but much higher in the sky. Do clouds only "light up" in space comics, or do they do it right here on Earth?

6. A real sucker

If you added water to a tornado, you'd end up with a swirling tube of water. Waterspouts start over lakes or rivers and can suck up fish as they whirl. Sometimes they drop the fish several kilometres (miles) away. Are waterspouts a fishy story, or do they happen right here on Earth?

WEATHER ON OTHER PLANETS

Have you ever experienced one of those days that starts out sunny, rains by noon, hails in the late afternoon and then ends just as it started? Weird! If you think the weather on Earth is weird, you should try living on some of the other planets.

Don't expect blue skies and sunshine on Mars. During the day the sky is peach coloured. But even though the sun is weak, it beams down lots of tanning ultraviolet rays. So Martians — if there are any — will have great tans. They probably won't spend a lot of time at the beach, though. The temperature on Mars is usually colder than at our North Pole.

Bad weather on Mars is *really* bad. Biting cold winds whip up surface dust, creating a frosty, dirty dust storm that blows so hard you'd have to cling on to something to stay on your feet.

But that's balmy compared to Triton, Neptune's largest moon. Icy slush squeezes out of cracks on Triton's surface, like a Slurpee gone crazy, and frozen lakes dot the landscape. Scientists believe it's the coldest place in our solar system.

Looking for somewhere warmer? Try Venus. It's so hot that a picnic lunch would vaporize as soon as you unpacked it. (So would you, unless you were protected by a spacesuit.)

You wouldn't need to bother with an umbrella on Venus. Although thick yellowish green clouds fill the sky, a rain made of sulphuric acid evaporates in the hot Venus air before it ever reaches the ground.

You might complain about the weather on Earth, but it doesn't take much planet hopping to see that there's no place like home. Why is Earth's weather so different from that of other moons and planets? The main reason is the Earth's distance from the sun. We're far enough away that we don't fry and close enough not to freeze.

The other reason is our atmosphere. This mixture of gases surrounds the Earth and acts like a blanket. It keeps in enough of the sun's warmth to allow us and the plants we eat to survive. But it blocks out most of the sun's harmful rays. Thanks to the sun and our atmosphere, Earth's weather is out of this world.

Pick a planet

Looking for somewhere to go on a holiday? Check out the temperature on some of our neighbouring planets before you buy a ticket.

Approximate daytime temperature

Mercury	410°C (770°F)
Venus	447°C (837°F)
Earth	14°C (57°F)
Mars	−125°C to 30°C (−193°F to 86°F)
Jupiter	−140°C to 24°C (−220°F to 75°F)
Saturn	−180°C (−292°F)
Uranus	−216°C (−357°F)
Neptune	−220°C (−364°F)
Pluto	−230°C (−382°F)

Aloha from Venus!

Picture Card

Dear Gang:
Having a wonderful time, wish you were here. Love, Jeff.

The Gang
585 ½ Bl
Toronto
Canad
Earth

BY SPACE MAIL

NO SUN, NO WEATHER

What would happen if someone flicked a switch and turned off the sun? It'd be dark, sure, but there would be other changes, too. Without the sun's warming rays, Earth would rapidly cool off. Then what? No one knows for certain. Maybe the winds would die down. After all, it's the sun's heating that creates the winds. And it might stop raining because there must be warmed rising air to produce clouds and rain. Before long, some scientists think, Earth's weather would grind to a halt.

69

FORECASTING THE WEATHER

From the time the first caveman licked his finger and held it up to find the wind direction, people have tried to forecast the weather. The equipment may be fancier, but modern meteorologists still have a tough time getting it right. Why is forecasting the weather so difficult? Read on to find out. Then test some weather sayings, make a weather predictor and find out about some weather that's so strange it would make a caveman run for his cave.

AND NOW FOR THE WEATHER FORECAST

What's sleepy and furry and forecasts the weather? The Easter Bunny one day after Easter? No, it's Punxsutawney Phil, a groundhog who gets national attention every February 2. When Phil crawls out of his burrow that morning, people gather to watch.

According to an old superstition, if Phil sees his shadow, Americans can expect six more weeks of winter. If he doesn't, spring is on the way. This is based on the old belief that a sunny day (when a groundhog can see his shadow) early in the winter means bad weather later.

The truth of the matter is that Phil's not a very accurate forecaster: he's wrong more often than he's right.

To be fair, Phil has a tough job: forecasting the weather six weeks ahead is very difficult. Meteorologists using sophisticated equipment can only forecast what the weather will be for about the next seven days.

Forecasts are based on information about such things as wind speed and direction, air pressure, moisture in the air, nearness of thunderstorms or other weather systems and temperature. Each bit of information is like a piece of a puzzle. By assembling these pieces, meteorologists can get a "picture" of the weather.

But the puzzle pieces keep changing, and some appear or disappear without warning. One day the wind will be blowing hard from the north; by the next day it might have switched to the south. A thunderstorm may suddenly build up, dump down rain and vanish. Imagine putting together a puzzle in which the pieces were always changing!

Thanks to satellites and other new ways of gathering information, meteorologists are getting better at forecasting all the time. Today's weather forecasts are almost twice as accurate as they were when your parents were kids.

WEATHER WATCHERS

The people and equipment on this page gather pieces of the weather puzzle and pass them along to meteorologists so that they can make their forecasts.

Cloud gun

Microwave beams from radar towers on the ground zap right through most clouds. But they can't pierce clouds that contain rain or snow. Information about the ways the beams bounce back gives meteorologists an advance warning of coming rain or snow.

Sky eyes

Weather satellites circle the earth or hang over the same place and beam back information on the movements of storms.

Human eyes

Staff at 100 000 weather stations around the world monitor the weather every hour. Some stations send up balloons equipped with measuring devices to find out what's happening in the air several kilometres (miles) up.

Volunteers across America measure the temperature twice a day and rainfall once or twice a day. The information they gather helps meteorologists understand the local climate.

BE A HUMAN THERMOMETER

Next time you're out in the country, listen for a field cricket and count the number of chirps you hear in eight seconds. Add four to the number of chirps and you'll have the temperature in Celsius. (For Fahrenheit, count the chirps in 15 seconds and add 37.) You'll find the cricket amazingly accurate, usually within one degree nine times out of ten. Crickets' ability to "chirp" the temperature has earned them the name "poor man's thermometer."

How good are *you* at telling the temperature? Try it and see.

You'll need:

- *a bowl of warm water*
- *a bowl of cold water*
- *a bowl of medium-temperature water*

1. Put your left hand in the bowl of cold water for a few seconds. Then put it in the medium bowl. Does the medium-temperature water feel warm or cool?

2. Put your right hand in the bowl of very warm water, then into the medium bowl. How does it feel?

3. Wait a minute or so, then put both hands directly into the medium bowl. Is it warm or cool?

It's a good thing you don't have to rely on your body to tell the temperature. You get confused by changes and your sense of hot and cold becomes mixed up. That's why scientists use instruments to measure the temperature and other weather conditions: instruments can't be fooled the way your body can.

You can tell a lot about the weather with just five instruments:

- a thermometer. It measures temperature. Now that you know how bad your body is at telling the temperature, you know why you need one.

- a rain gauge. It tells you how much rain has fallen. See page 33 for how to make one.

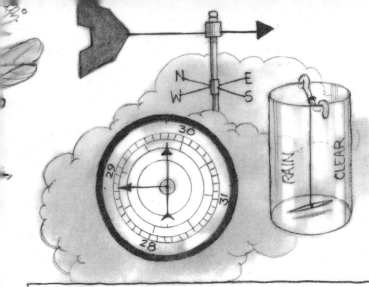

- a wind vane. It tells what direction the wind is blowing from. Make one on page 45.

- a hair hygrometer. It measures the amount of moisture in the air. Read more about it on page 34.

- a barometer. It measures air pressure: increasing pressure usually signals good weather, while sinking pressure means you're in for bad weather. To make a barometer, see the box on this page.

MAKE A SIMPLE BAROMETER

This home-made barometer will show you the changes in air pressure that herald good or bad weather.

You'll need:

○ *a balloon*
○ *a jar with a wide mouth*
○ *an elastic band*
○ *a straw*
○ *tape*
○ *a piece of paper and a pen*

1. Blow up the balloon, then let out the air. Cut the balloon in half and throw the part with the neck away. Stretch the other part over the mouth of the jar and attach it with an elastic band.
2. Tape a straw in place as shown.

3. Set your barometer next to a wall and tape a piece of paper up beside it.
4. Mark the position of the straw for several days. What kind of weather makes the straw point highest? Lowest? Now you'll know what kind of weather to expect by watching how high the straw points.

How does your balloon barometer work? Air pressing down on the stretched balloon makes it curve down, which forces the free end of the straw up.

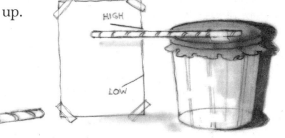

The more pressure there is pushing down on the balloon, the higher the straw points. And the better the weather.

LAMBS AND LIONS

Have you heard this old saying: "If March comes in like a lion, it'll go out like a lamb"? Long before there were meteorologists or weather satellites or other high-tech equipment, people relied on weather sayings like this one to forecast the weather.

For fishermen and farmers, being able to predict the weather could mean the difference between life and death. If farmers waited too long to harvest their crops in the fall, rain could turn their fields into mush. Fishermen who ignored storm signals might find themselves awash.

How accurate were the old weather sayings? Why not do a test and see? Photocopy the chart on the opposite page. Pick one or more of the weather expressions below, then fill in the chart and test its accuracy. (To make your test fair, record the temperature at the same time every day, such as when you get home from school. To measure the rainfall, you can make your own rain gauge using the instructions on page 33.)

1. *If March comes in like a lion it'll go out like a lamb.* If the weather is ferocious at the beginning of March, there'll be mild weather at the end of the month.

2. *Red sky at night, sailor's delight; red sky in the morning, sailor's warning.* A pinkish-red sky in the evening will mean fine weather the next day; the same colours in the morning are a storm warning.

3. *The first three days of January rule the weather for the next three months.* The weather on the first three days of January foretells the weather for the next three months.

Just for fun, why not look at the long-range weather forecast in the newspaper and do the same test on it!

MY WEATHER CHART

DATE	TEMPERATURE	MILLIMETRES (INCHES) OF RAIN OR SNOW	CLOUDS	WINDS (strong or mild)

A SNOWY DAY IN JULY

Suppose you overhear your mother talking on the telephone and she says, "It'll be a snowy day in July before I'll drive the kids to school again." Better be prepared to start walking.

Without even thinking about it, you have come to expect certain kinds of weather at certain times of the year. And snowy days in July just don't fit in.

Your idea of what's normal depends on the climate in your area. For example, if you live in Hawaii, you probably think of January as beach weather. In most parts of Canada you'd have to break through the ice to go for a swim in January. Both are normal.

Each place has its own climate controls, like a computer program, that keep the climate pretty much the same year after year. The most important climate control is how far north or south you live. The closer to the equator, the more of the sun's warmth you get. That's why you can take a January swim in Hawaii, but not in Hudson Bay.

Do you live near the ocean or a large lake? Your climate may be cloudier and rainier than that of places farther inland. Your summers will probably be cooler and your winters warmer, too. If you're landlocked, smack in the middle of a continent, the temperature where you live may go up and down more in a single day and from one day to the next than if you live near a large lake. Your summers are likely to be hot and your winters cold. Each area has its own climate controls that shape what the weather is like.

Weather can change overnight, but climate stays the same for many hundreds of years because the same controls are "programming" it. But sometimes the controls change, and when that happens, so does the climate. The result? Palm trees in Canada and, yes, snowy days in July. Turn the page to find out more about Earth's climate long ago.

SOME COLD AND SNOWY JULY DAYS

July is right in the middle of winter for New Zealand and other Southern Hemisphere countries, so snow does sometimes fall in July. The farther south you go in July, the colder it gets. The coldest temperature ever recorded was on July 21, 1983, way down south in Antarctica.

Even the Northern Hemisphere sometimes gets snow in July. British explorer Martin Frobisher recorded a full-fledged blizzard in July 1578 in Northern Canada. He wrote in his journal: "In this storm, there fell much snow with such bitter cold air that we could scarce see one another. . . . The snow was about half a foot deep." The men who travelled with Frobisher were amazed: if it snowed in the summers, what would the winters be like, they wondered.

CAMELS IN THE ARCTIC?

When scientists dug up the fossilized bones of a camel in the Arctic, they could hardly believe their eyes. What was an animal that usually lives in hot desert areas doing in the cold Canadian North?

More camel bones were found, along with plant remains. These clues hinted that the Arctic wasn't always as cold and barren as it is today. Thousands of years ago, Arctic summers seem to have been longer — long enough for plants to grow and feed the camels. And winters were probably less snowy. The camels must have been able to find food and survive — something today's camels couldn't do in the North.

If you could travel back in time millions of years, you'd go through several warm periods in Earth's climate. Average temperatures around the world rose only a few degrees, but the warming was enough for camels to live in the Arctic and, even earlier, dinosaurs in Alaska.

Between these warm periods were the ice ages. Just how icy were the ice ages? Very.

Clue no. 1: tree rings. Trees in cold climates grow new layers each year. Counting the rings from the outside layer in is like going back in time, year by year. Since trees usually grow more in wet years, scientists can often tell whether a year was dry or had lots of rainfall by studying the thickness of the rings.

Clue no. 2: plant pollen that fell into lakes long ago and fossils. Scientists compare these plant remains with modern plants to solve mysteries about the climate. For example, if they find a fossil of a palm-like plant, they know the climate must have been warm.

Clue no. 3: preserved animal remains, such as fossils. One particularly helpful fossil is the *Globigerina pachyderma*. The shell of this tiny sea creature can be read like a thermometer. The Globi's shell twirled to the left when it was cold and to the right when it was warm.

These clues tell scientists that Earth's climate has warmed and cooled over and over again. But no one knows for certain why. Scientists do know that it's the amount of heat from the sun that controls Earth's "temperature." They believe that this heat might have been "turned down" or "turned up." How? A thick cloud of ash and gas from a volcano's eruption may have blocked out the sun's heat. Or the earth's irregular orbit may have taken it farther from the sun from time to time. A combination of Earth's irregular orbit, tilt and wobble would have had a great impact on the earth's climate.

Could we Earthlings be in for another ice age? Looking back over the weather for the last few hundred years, scientists believe we're in a cooling period. But the greenhouse effect (see page 54) is warming Earth, so no one can be sure if our planet will get warmer or cooler during the next few decades. One thing *is* sure: you're not likely to see camels in the Arctic — unless a circus is visiting.

Temperatures dropped, snow mounted and glaciers crept over much of North America and Europe. The last big glacier push, called the Wisconsinian, peaked about 20 000 years ago. Check the map to see what was under ice.

Scientists who study the Earth's past climate are like detectives. All they've got to go on are clues left behind by long dead plants and animals:

81

WEATHER WHYS

Why are rain clouds so dark? How acid is acid rain? Who is Jack Frost? Read on for the answers to these and other weather whys, whats, whos and hows.

Q. What's a weather map?

A. It's a map that shows the main high and low pressure areas across the country and the cold and warm fronts. The weather map in your local paper might look like the one on the left. H stands for a high-pressure area, which generally means good weather; L stands for a low-pressure area, which often spells bad weather. The fronts (the edges of warm or cold air masses) are marked like this:

WARM FRONT   COLD FRONT

Some weather maps use symbols to give other information. Here are some of the international weather symbols. If you don't find them on the weather map in your newspaper, try marking them on yourself.

HAZE

DRIZZLE

RAIN

SHOWERS

THUNDERSTORM

LIGHTNING

FREEZING RAIN

SNOW

Weather maps can tell you what kind of weather to expect. Weather moves from *west to east*. So if you want to know what kind of weather is on its way, look at what's happening on the map west of where you live.

Q. How acid is acid rain?

A. You don't need to worry about it eating holes in your umbrella. It's not that acidic. But acid rain that falls into lakes can harm the lakes and kill the fish.

The acidity of rain is measured by its pH level. Pure rain has a pH level of 5.6. Acid rain is any rain with a pH level *lower* than 5.6.

If too much acid rain falls into a lake, the lake may become acidified. You can see what happens to the fish as the pH level drops. Fish cannot survive with a pH level below 4.5.

Q. What is windchill?

A. Ever noticed how you feel colder when the wind blows? That's because the wind blows away a thin layer of warm air that usually surrounds your body. On windless days this layer of air acts like insulation in a house and helps to keep you warm; on windy days — whoosh! — it's whipped away and you lose body heat.

A scientist named Paul Siple came up with a way of figuring out just how much colder you would feel with various speeds of wind. For example, if the thermometer said it was −18°C (0°F) and the wind was blowing at 16 km/h (10 mph), he calculated that it would feel more like −30°C (−22°F). He called this windchill. Today weather forecasters often announce the windchill so that people will know how to dress for the cold.

pH HIGHER THAN 5.0

pH 4.3 TO 5.0

Q. What's a chinook?

A. It's a hot, dry wind that can turn winter into summer in a matter of minutes. One blustery cold winter day in 1943, the townspeople of Rapid City, South Dakota, went from snowsuits to shorts in just 15 minutes as a sudden chinook boosted the town's temperature from −12°C (10°F) to 13°C (55°F).

Where does such a hot dry wind come from? A moisture-laden wind flows up one side of a mountain and dumps its moisture. The now-dry wind warms as it flows down the other side of the mountain.

The chinook got its name from an Indian word meaning snow-eater. No wonder. A chinook wind can "eat" a knee-deep snowfall overnight. That sure beats shovelling the stuff.

Q. What's El Niño?

A. It's a warming of the ocean's surface that happens around Christmas every year off the coast of Ecuador and Peru. Some years it's hardly noticeable. But other years, El Niño turns the world's weather upside-down. Places that are normally sunny and dry, such as Florida, are drenched with rainstorms. Countries expecting a much-needed rainy season, such as Indonesia, have a drought. All of this happens because the warm ocean water of El Niño starts a chain reaction in the atmosphere and shakes up the normal weather patterns.

Q. Why are rain clouds so dark?

A. Because they're loaded with ice crystals, cloud droplets and rain droplets that are about to fall. The droplets and crystals block the sunlight from coming through the cloud. The more droplets and crystals, the less light that gets through, and the darker the cloud looks. Really dark clouds contain a lot of snow, which blocks the light even more.

pH LOWER THAN 4.2

Q. Who is Jack Frost?

A. Jack Frost came from a pretty cool family. His father was a Scandinavian wind god named Kari. Jack himself had a son named Snjo, or Snow in English.

It was Jack's job to decorate the world with frost in winter. He did this fairly well for hundreds of years until (horrors!) the real story of frost was discovered by scientists. Jack was out of a job.

You can make frost just as Jack Frost did.

You'll need:

o *ice cubes*
o *a plastic bag*
o *a hammer*
o *an empty tin can*
o *salt*

1. Put the ice cubes into a plastic bag and tie the end shut. Break the ice cubes into small chips with the hammer.

2. Put a layer of ice in the tin can about 3 cm (1 inch) thick, then a thin layer of salt, another layer of ice and so on until the can is full.

3. Watch what forms. If it's a dry day, with not much humidity in the air, you may have to breathe on the outside of the can to produce frost.

What happened? When moist air hits the cold surface of the can, the water in it freezes and sticks to the can. The result is frost. This process of going from a gas (your breath) to a solid (frost) is called sublimation. It's what happens when frost forms in winter.

Q. Can you can smell rain before it falls?
A. Some scientists think you can. They believe that plants give off more scent when the air is moist, before a rainfall. When people smell these scents, they have learned to expect rain.

Other scientists believe it's all in your nose. They say the moisture in the air tickles your sense of smell and makes it more sensitive. Try this for yourself. Compare how strong a lemon smells in a misty bathroom and in a dry room.

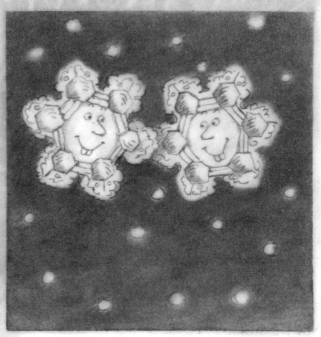

Q. Is it true that no two snowflakes are alike?
A. That's what people have believed for a long time. And for good reason. To be identical, two snowflakes would have to fall through *exactly* the same conditions. The tiniest extra bit of moisture or the smallest puff of wind will make two flakes look different.

Identical snowflakes seemed impossible — until physicist Nancy Wright caught two look-alikes when she was collecting snowflake samples in the winter of 1988. She took pictures of the snowflakes and enlarged them. To the naked eye, they appear to be identical. But even so, Nancy Wright isn't convinced. She says there may be differences that are so tiny we can't spot them.

OUCH!!

A HAILSTONE BIGGER THAN A GRAPEFRUIT CRASHED DOWN IN POTTER, NEBRASKA, ON JULY 6, 1928. BUT THE HEAVIEST HAILSTONE ON RECORD FELL IN BANGLADESH ON APRIL 14, 1986. IT WEIGHED 1 KG (2¼ POUNDS)

SWEAT and SHIVER

In a single day, the temperature difference between the North and South Poles can be as much as 82°C (180°F).

IT'S A RECORD

Hang on to your hat, get out your umbrella and don't forget your bathing suit. Here's some of the windiest, wettest, hottest and downright weirdest weather anywhere on earth.

RAINED OUT

July 4, 1956: 31 mm (1.22 inches) of rain fell in Unionville, Maryland, in one minute. If it had rained that hard for an hour, the water would have been over peoples' heads.

GASP!!

ARICA, CHILE, ONCE WENT FOR 14 YEARS STRAIGHT WITHOUT RAIN.

HOLD ON TO YOUR HAT !!!

Winds have been measured at 303 k/h (188 mph) near Jan Mayen Island. That's fast enough to rip your clothes off.

ZZAP!!

GOR, INDONESIA, ONCE HAD 2 DAYS OF THUNDERSTORMS A YEAR. THAT'S MORE THAN 10 MONTHS WORTH OF THUNDERSTORMS.

SPLAT!!

A giant raindrop 8 mm (⅓ inch) across was found near Hilo, Hawaii. The world's biggest snowflake was bigger than a long-playing record—38 cm (15 inches) in diameter.

HHMMMM

The average world temperature is 15°C (59°F).

START SHOVELLING!!

The winter of 1955–56, Mt. Rainier, Washington, was buried under 25.4 m (83 feet) of snow, enough to cover a four-storey apartment building.

WHEW!!! AND BRRRR

HOT ENOUGH FOR YA?

The hottest spot on Earth is Al'Aziziyah, Libya. On September 13, 1922 the temperature hit 58°C (136.4°F). The cold spot is Vostok, Anarctica, where the temperature has dipped to −88.3°C (126.9°F).

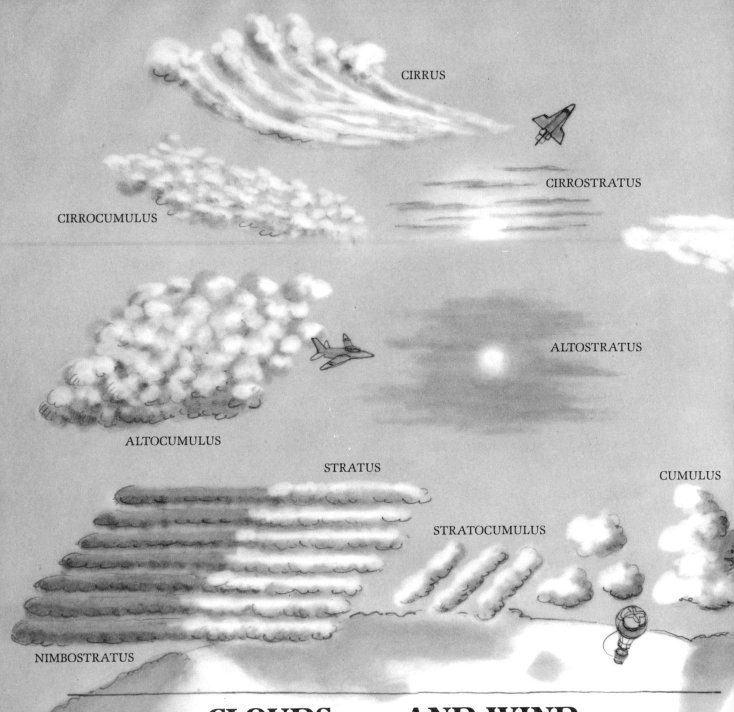

CIRRUS

CIRROSTRATUS

CIRROCUMULUS

ALTOSTRATUS

ALTOCUMULUS

STRATUS

CUMULUS

STRATOCUMULUS

NIMBOSTRATUS

CLOUDS . . . AND WIND

Here are the main kinds of clouds. You'll never see all 10 all at once, but you might see two or three.

THE BEAUFORT WIND SCALE

BEAUFORT NUMBER	KM/H	MPH	DESCRIPTION
0	less than 1		Calm: smoke rises vertically
1	1–5	1–3	Light air: not enough to move a wind vane, but shows in smoke drift
2	6–11	4–7	Light breeze: wind felt on face, leaves rustle, wind vanes moved by wind
3	12–19	8–12	Gentle breeze: leaves and small twigs in constant motion, enough to blow a light flag
4	20–29	13–18	Moderate: raises dust and loose paper, moves small branches
5	30–39	19–24	Fresh: small trees in leaf begin to sway, crested wavelets form on inland waters
6	40–50	25–31	Strong: large branches move, umbrellas used with difficulty
7	51–61	32–38	Near gale: whole trees in motion, inconvenience in walking against the wind
8	62–74	39–46	Gale: wind breaks twigs off trees, makes walking difficult
9	75–87	47–54	Strong gale: slight damage occurs to building structures
10	88–102	55–63	Storm: trees uprooted, considerable damage
11	103–121	64–72	Violent storm: widespread damage
12	over 122	over 73	Hurricane

CUMULONIMBUS

CREDITS

The sun separator, page 37, reprinted from OWL
Magazine with permission of the publisher, The Young
Naturalist Foundation.

The Flying Meat Tray, page 53, from *Super Flyers*. Text ©
1988 by Neil Francis. Reprinted by permission
of Addison-Wesley Publishing Company.

ANSWERS

Make a rainbow in your backyard,
page 36:
Sun shining through a fog can produce a fogbow.

Snow what!, page 41:
1. True—it was probably coloured pink by algae in it; 2. b; 3. false; 4. the largest snowflake measured 38 cm (15 inches) across; 5.b.

Winds on the move, page 47:
Elephanta, India; haboob, Sudan; zonda, Argentina; williwaw, Alaska; xlokk, Malta; simoom, North Africa; kwat, China; bhoot, India.

It came from outer space, pages 66–67
It may sound as if all these weather phenomena belong in outer space but they actually happen right here on earth. **1.** You can sometimes see *virga* when the clouds are high and the air is dry. Virga evaporates long before it hits the ground so you wouldn't feel anything if you stood under a virga cloud. **2.** *Ball lightning* is an extremely rare form of lightning that looks like a slow-moving sphere. **3.** Yes, *snowrollers* happen, but not very often. Some scientists think they may start out as solid rolls, then the middle evaporates, leaving a hole. **4.** *Herbies* happen, but only in Antarctica. Meteorologists stationed there have learned to watch the Weddell seals for advance warning. When the seals disappear through holes in the ice, a herbie's about to hit. **5.** *Glow-in-the-dark, or "noctilucent," clouds* happen when clouds of tiny dust particles high up in the sky are illuminated by the sun, which is below the horizon. **6.** *Waterspouts* are responsible for some unusual "rains." On October 23, 1947, fish, some as long as a pencil, "rained" down on the town of Marksville, Louisiana. Montrealers experienced a rain of live lizards in 1857. Residents of the town of Worcester, England, were astonished to see crabs and periwinkles (a kind of snail) rain down in 1881. All these raining animals were probably due to waterspouts.

INDEX